Healthy Recipes for Chronic Kidney Patients & Their Families

By Barbara Clark & Robert Minard

PublishAmerica
Baltimore

© 2008 by Barbara Clark and Robert Minard.
All rights reserved. No part of this book may be reproduced, stored in a retrieval system or transmitted in any form or by any means without the prior written permission of the publishers, except by a reviewer who may quote brief passages in a review to be printed in a newspaper, magazine or journal.

First printing

PublishAmerica has allowed this work to remain exactly as the author intended, verbatim, without editorial input.

ISBN: 1-60563-145-0
PUBLISHED BY PUBLISHAMERICA, LLLP
www.publishamerica.com
Baltimore

Printed in the United States of America

Contents

Cheesy Tuna Ball 7
Crab Spread 8
Cocktail Spread 9
Horseradish Deviled Eggs 10
Horseradish & Cheese Dip 11
Salmon Spread 12
3 Pepper Appetizers 13
Tropical Fruit Dip 14
Cabbage Soup 15
Chicken and Rice Soup 16
Garden Vegetable Soup 17
Turkey Soup 18
Apple Fruit Salad 19
Chicken Salad 20
Crab Salad 21
Egg Salad 22
Lettuce Salad with Fruit 23
Radish Salad 24
Cream Herb Dressing 25
Tropicana Gelatin 26
Vegetable Gelatin Salad 27
Barb's Dressing 28
Citrus Dressing 29
Garlic Mayo Dressing 30
Herb Dressing 31
Herbed Vinaigrette 32
Mint Vinaigrette 33

Italian Dressing	34
Mock Hollandaise Sauce	35
Oriental Dressing	36
Raspberry Honey Vinaigrette	37
Tangy Dressing	38
Baked French Toast	39
Egg Omelet with Biscuits	40
Fat Free Apple Waffles	41
Oatmeal Breakfast Bars	42
Pancakes	43
Potato Pancakes	44
Power House French Toast	45
Strawberry Cream Cheese French Toast	46
Zucchini Pancakes	47
Baked Hash Browns	48
Baked Zucchini or Egg Plant	49
Baked Green Beans and Onions	50
Honey and Cranberry Acorn Squash	51
Roasted Italian Spaghetti Squash	52
Pearl Onion Side Dish	53
Roasted Garlic Mashed Potatoes	54
Roasted Baby Carrots and Rutabaga	55
Oriental Peas	56
Roasted Three Pepper Zucchini	57
Refrigerator Pickles	58
Apple Stuffed Pork Chops	59
Braised Beef or Pork Short Ribs	60
Broiled Steak	61
Chicken and Biscuit Casserole	62
Chicken Patties	63
Easy Parmesan Chicken Breasts	64
Florentine Pork Chops	65
Hawaiian Meatballs	66
Honey Mustard Chicken	67

Italian Sausage with Pasta	68
Oriental Glazed Chicken Bites	69
Pina Colada Chicken	70
Ravioli in Garlic Butter Sauce	71
Rice Stuffed Flounder	72
Roasted Chicken Sausage	73
Seared Pork Loin and Vegetables	74
Seared Tuna and Bow Ties	75
Shrimp with Mixed Peppers	76
Tuna Bake	77
Turkey with Fruit Stuffing	78
Swedish Meatballs and Noodles	79
Apple Butter Cake	80
Apple Cake	81
Apple Pie Delight	82
Apple Streusel	83
Black Cherry-Vanilla Italian Ice	84
Cheese Cake	85
Cranberry-Maple Pie	86
Flan	87
Frozen Pudding Cream Pie	88
Pear Crisp	89
Pineapple or Applesauce Spice Cake	90
Quick Strawberry Cheese Cake	91
Upside Down Pear Cake	92

Cheesy Tuna Ball

1 (8 oz) package cream cheese
1 (6 oz) can tuna
2 teaspoons lemon juice
1 teaspoon garlic powder
1/4 teaspoon horseradish
1/4 cup parsley

With electric mixer combine softened cheese with all but parsley. Mix well shape into a ball then roll in parsley. Chill well and serve with crackers.

Crab Spread

1/4 cup liquid non dairy creamer
8 oz low fat cream cheese
2 teaspoons (bottled) lemon juice
1 teaspoon (bottled) minced garlic
1 cup cooked-cut up crab meat or imitation crab meat

Mix all in medium mixing bowl with Electric mixer. Do not over mix. Serve with unsalted crackers.

Cocktail Spread

1 pkg. (8 oz) cream cheese
1 small amount of mayonnaise
1 (8 oz) frozen bag shrimp salad

Soften cream cheese & spread on plate Pour cocktail sauce ever the cream cheese to desired thickness. Serve with crackers.

Horseradish Deviled Eggs

6 hard boiled eggs
1/4 cup mayonnaise
1 teaspoon horseradish
1/2 teaspoon dill weed
1/4 teaspoon ground mustard
dash paprika

Cut eggs lengthwise, remove yolks. In bowl mash yolks and mix all ingredients. Spoon the mixture into the egg whites. Sprinkle with paprika. Refrigerate.

Horseradish & Cheese Dip

6 hard boiled egg
1/4 cup mayonnaise
1 Tablespoon horseradish
1/2 tsp dill weed
1/4 teaspoon ground mustard
Dash of paprika
1 (8 oz) low fat cream cheese

Cut egg lengthwise, remove yolks. In bowl mash yolks and mix all ingredients. Spoon the mixture into the egg whites. Sprinkle with paprika. Refrigerate.

Salmon Spread

1/4 cup non fat mayonnaise
1 teaspoon white vinegar
1 1/2 cups cooked salmon
1 Tablespoon minced onion
1 teaspoon lemon juice

Mix all ingredients in mixing bowl using electric mixer.

3 Pepper Appetizers

3 peppers, (1 yellow, 1 red, 1 green)
1 (8 oz) fat free cream cheese
1 1/2 teaspoons lemon pepper spice
1/2 teaspoon bottled lemon juice
1/4 teaspoon powdered garlic
1/4 cup toasted coconut

Wash & cut peppers in half, remove seeds. Cut into bit size bit sized pieces. Place rest of ingredients in a small mixing bowl (except coconut). Using electric mixer, mix until creamy. Drop the cream cheese mixture on each pepper. Sprinkle with toasted coconut. Refrigerate.

Tropical Fruit Dip

1 (18 oz) low fat cream cheese softened
1 (small package) coconut cream pudding
2 Tablespoons toasted coconut
1 small can mandarin oranges drained—except 2 Tablespoons
1 (8 oz) can crushed pineapple drained

Mix all ingredients in large mixing bowl with electric mixer. Refrigerate 2 hours before serving.

Cabbage Soup

2 cups reduced sodium chicken broth
1 cup filtered water
1 small cabbage shredded
1 small can carrots (low sodium)
1/4 cup soup noodles

Place all ingredients in a large pot (except carrots). Heat to a boil, turn heat down and simmer for 15 minutes. Add carrots.

Chicken and Rice Soup

3 cups filtered water
3 cups chicken stock or reduced sodium chicken broth
3 carrots sliced, or low sodium canned drained
1 cup rice
One can green beans low sodium, or fresh or frozen
1/8 teaspoon garlic powder
1/2 teaspoon parsley
1 1/2 cups cooked chicken

Heat the water in large pot. Add broth and spices. When water is boiling add rice and raw vegetables. If using canned vegetables wait then add with chicken to the cooked rice. Cook rice 20 minutes.

Garden Vegetable Soup

1 small zucchini diced
1 cup onions diced, fresh or frozen
1 cup carrots, canned low sodium drained or fresh
1 cup green beans, canned low sodium drained or fresh
2 cups low sodium chicken broth
1 cup filter water
1/4 cup noodles of choice

Place all ingredients in large pot. Heat at a simmer until vegetables is tender. If using canned vegetables, add after fresh vegetables are cooked. Add noodles and bring to a boil until noodles are done.

Turkey Soup

1 cup cooked turkey-diced
3 cups filter water
3 cups reduced sodium chicken broth
1/2 cup rice
2 medium onions chopped
1 (10 oz) package frozen corn
1/2 oz package frozen peas
3 medium carrots sliced
2 pieces celery sliced
1/2 teaspoon garlic power (optional)

In large pot combine turkey, water, broth and rice, simmer covered for 30 minutes. Add onions, corn, peas, and carrots, simmer covered 30 minutes or until vegetables are tender.

Apple Fruit Salad

1 cup peeled chopped sweet eating apple
1 cup seedless grapes, red, purple or black washed and cut in half.
1 1/2 teaspoon real lemon juice
1/2 cup low fat mayonnaise or sugarless whipped topping
1/4 teaspoon toasted coconut (optional)

Place apple and grapes in a medium sized bowel, toss with lemon juice. Add mayonnaise or whipped cream. Top with coconut. Refrigerate for one hour and 30 minutes before serving.

Chicken Salad

1 cup cooked chicken diced
1/2 cup nonfat mayonnaise
1/8 cup green pepper diced
1/8 cup celery sliced
1 teaspoon horseradish

Mix all ingredients in bowl.

Crab Salad

4 oz low fat cream cheese softened
1/4 cup herbed vinaigrette
1 1/2 teaspoons lemon juice
1 teaspoon bottled mixed garlic
1/8 teaspoon course black pepper
1 teaspoon chopped parsley
1 pkg. imitation crab diced

Mix all ingredients and refrigerate. Serve with vegetables or unsalted crackers.

Egg Salad

4 hard boiled eggs
1/2 cup nonfat mayonnaise
1 teaspoon brown mustard
1/8 cup celery sliced

Mix in medium bowl.

Lettuce Salad with Fruit

4 cups washed and shredded lettuce
1 teaspoon olive oil or canola oil
1/4 cup unseasoned salad croutons
1 Tablespoon white vinegar
3 Tablespoons whole canned cranberry sauce
Parmesan cheese (optional)

Place lettuce in large bowl. Mix oil, vinegar, and add croutons and cranberry sauce in small cup. Pour on lettuce and add croutons.
Or use any allowed fruit with 3 Tablespoons reduced sugar or fruit preserves.

Radish Salad

Radishes washed and thinly sliced
5 pearl onions (optional)
2 tablespoons Italian dressing
1/2 teaspoon parsley

Add all ingredients in bowl, stir. Serve at room temperature.

Cream Herb Dressing

1/2 cup non dairy creamer
1/4 cup roasted garlic rice vinegar
1/2 teaspoon dry minced or 1/4 cup fresh minced
1/2 teaspoon basil (optional 1 Tablespoon)

Place all ingredients in air tight container and shake. Refrigerate

Tropicana Gelatin

2 teaspoons rum extract
1 (6 oz pkg.) sugarless orange gelatin
1 can Mandarin oranges drained
1 cup filtered boiling water
1 cup Piña Colada mix
1/2 cup coconut (shredded)
1/2 jar Manichino Cherries (halved)

Mix water and gelatin until dissolved. Add remaining ingredients, stir well. Refrigerate.

Verifications:
Add Marshmallows.
Use 1 can fruit cocktail drained in place of Mandarin oranges.

Vegetable Gelatin Salad

1 large pkg. sugar free lime gelatin
1/2 cup shredded carrots
1/2 cup shredded cabbage
1/4 cup low fat cottage cheese (optional)

Prepare gelatin as directed on package (use filtered water). Add all ingredients when gelatin is half way set, stir. Continue to refrigerate until set.

Barb's Dressing

1/4 cup sugar or sugar substitute
3 Tablespoons lemon juice
1 cup olive oil
1 Tablespoon Worcestershire sauce
1 Tablespoon minced dry onion
1 Tablespoon white vinegar
1 1/2 teaspoon celery seed

Place all ingredients in an airtight container and shake. Refrigerate.

Citrus Dressing

2 small tangerines peeled, membrane removed
4 Tablespoons white vinegar
3 green onions minced
1/2 teaspoon garlic powder
2 Tablespoons sugar or sugar substitute
1/2 cup olive oil
2 Tablespoons bottled lime juice
1/8 teaspoon course black pepper

Place all ingredients in blender. Whip until fruit is well chopped.

Garlic Mayo Dressing

1/2 cup low fat mayonnaise
1/2 teaspoon garlic powder
1/4 teaspoon dry minced onion
1 teaspoon ground basil

Place in small bowl and stir well. Chill in airtight container.

Herb Dressing

1/2 cup low fat mayonnaise
1/4 cup roasted garlic rice vinegar
1/2 teaspoon dry minced onion
1/2 teaspoon basil
1/4 teaspoon cilantro
¼ teaspoon parsley
1 Tablespoon sugar or sugar substitute
1 teaspoon brown mustard

Place in small bowl and stir. Place in airtight container and chill.

Herbed Vinaigrette

1/4 cup olive oil
1/4 cup roasted garlic rice vinegar
1/2 teaspoon oregano
1/2 teaspoon basil
1 Tablespoon minced onion

Place in airtight container and shake. Refrigerate.

Mint Vinaigrette

1/2 cup olive oil
1/4 cup white vinegar
1 teaspoon honey
1/8 teaspoon course black pepper
1 Tablespoon minced mint leaves or 3 drops mint extract

Place in airtight container and shake. Refrigerate.

Italian Dressing

5 Tablespoon red or white vinegar
1/4 cup filtered water
1/4 cup olive oil
1 teaspoon sugar or sugar substitute
1/8 teaspoon course black pepper
1 teaspoon cilantro
1 teaspoon garlic

Place in airtight container and shake well. Refrigerate.

Mock Hollandaise Sauce

1/8 teaspoon cream cheese
1/4 cup low fat mayonnaise
1 teaspoon bottled lemon juice
1/2 teaspoon brown mustard

Place in small bowl. Stir well. Refrigerate in airtight container.

Oriental Dressing

2 teaspoons Chinese five spice
1 1/2 Tablespoons sugar or sugar substitute
1/2 cup olive oil
1/4 to 1/2 garlic roasted rice vinegar to taste
1 teaspoon bottled crushed ginger

Place in airtight container and shake. Refrigerate.

Raspberry Honey Vinaigrette

1/2 cup olive oil
1/2 cup apple juice
1 teaspoon dry minced onion
1 teaspoon course black pepper
1 Tablespoon honey
1 teaspoon oregano
3 Tablespoons red raspberry preserves

Place in airtight container and shake. Refrigerate.

Tangy Dressing

1 cup filtered water
1/3 cup apple cider vinegar
1 teaspoon dry minced onion
3/4 cup brown sugar or sugar substitute
1 teaspoon brown mustard
1 Tablespoon celery seed
2 Tablespoons cornstarch

Place all in small saucepan. Bring to a boil until thickened. Chill in airtight container.

Baked French Toast

8 pieces stale white bread
4 eggs or egg substitute
2 Tablespoon soft zero trans fat margarine melted in microwave
1/2 cup liquid non dairy creamer
1/4 cup sugar or sugar substitute
1 teaspoon cinnamon

Spray 8x8 pan with olive oil spray. Cube bread by stacking and cutting into strips. Then cut in strips the opposite direction. Place in pan. In small bowl mix the rest of the ingredients with a whisk. Pour over bread. Push bread down in egg mixture. Bake 20-30 minutes at 350 F. Dust with powdered sugar or fruit syrup.

Egg Omelet with Biscuits

Biscuit mix:

2 envelopes of rapid rise yeast
1/2 teaspoon sea salt (needed for yeast to rise)
3/4 cup filtered water
1 1/2 cups unbleached flour
3 Tablespoons olive or canola oil

Spray an 8x8 baking pan with olive oil. Mix all ingredients in pan.

In small bowl add the following and whisk:
4 eggs or egg substitute
1/2 teaspoon basil
1/2 teaspoon garlic
1/8 teaspoon black pepper
3/4 cup frozen diced green pepper and onion

Sprinkle 1/2 cup imitation cheddar cheese over biscuit mix. Pour egg mix over cheese. Bake in non pre heated oven at 350 F for 30 minutes or until knife inserted in middle comes out clean.

Fat Free Apple Waffles

4 eggs or egg substitute
2 cups liquid non dairy creamer
1 1/2 Tablespoon apple sauce
1/2 teaspoon cinnamon
2 cups unbleached flour
1 1/4 teaspoons baking powder
1/2 teaspoon baking soda

Mix all ingredients. Pour by spoonfuls on hot waffle maker. Waffles are done when no steam is coming from it.

Oatmeal Breakfast Bars

1 3/4 cups old fashion oatmeal
1 1/2 cups unbleached flour
1/2 cup brown sugar substitute
1/2 teaspoon baking powder
1/4 cup corn syrup
1/2 cup olive or canola oil
1 cup sugar free preserve jam

In medium mixing bowl put in all ingredients except preserves. Mix together with electric mixer, starting on low. Spray 11 x7 baking pan. Press 1/2 of mixture in pan. Bake at 350 F for 7 minutes. Remove from oven and spread with preserves. Crumble remaining oat mixture on top. Replace in oven for 20 minutes. Cool and cut into bars.

Pancakes

1 1/4 cup non dairy creamer
1 egg or egg substitute
2 Tablespoons olive oil or canola oil
2 Tablespoon zero trans fat unsalted margarine
2 1/2 Tablespoons sugar or sugar substitute
1 cup unbleached flour
2 teaspoon baking powder

Place egg, creamer, oil and sugar in medium bowl. Mix with electric mixer. Add rest of ingredients and thoroughly mix. Pour on lightly greased pan at 350 F. Turn once when bubbles break on pancake.

Potato Pancakes

2 cups leftover soaked mash potatoes
1 egg or egg substitute
1/4 cup minced onion
1/4 cup minced green pepper
1 Tablespoon olive oil
1/2 teaspoon baking powder
2 teaspoons unbleached flour

Mix all ingredients in medium mixing bowl, mix with electric mixer on low until well combined. Add a little liquid non-dairy creamer, if mixture is too stiff. Heat oil in skillet at 300 F. Drop by tablespoons to make patties. Fry until just crisp. Turn. Drain on plate with paper towel.

Power House French Toast

6 pieces white bread
4 eggs
2 Tablespoons honey
1 cup non dairy creamer
1 teaspoon rum extract
1 1/2 cups boxed cornflake crumbs
4 tablespoon soft zero trans fat margarine
3 Tablespoon cinnamon
3 Tablespoons sugar or equitant equivalent amount of sugar substitute
4 teaspoon baking soda

In shallow baking pan mix eggs, creamer, honey, rum and baking powder. In another shallow baking pan place cornflake crumbs. Dip bread in egg mix and turn over. Place bread in cornflakes and turn over. Place in electric fry pan with melted margarine. Cook until crisp. Mix sugar and cinnamon together and sprinkle over each piece. Can also use syrup of choice or fruit spread.

Strawberry Cream Cheese French Toast

8 pieces stale white bread
3 1/4 cups liquid non dairy creamer
2 eggs or egg substitute
1 1/2 teaspoons vanilla extract
8 Tablespoons cream cheese with fruit blend

In 8x8 pan mix all ingredients except bread and cream cheese. Dip bread quickly in liquid and place in sprayed skillet at 375 F, repeat until all bread is used. Turn once. Spread 1 Tablespoon of cream cheese mix on top. Cover and cook until browned on bottom. Serve with syrup or fruit syrups.

Zucchini Pancakes

2 cups shredded zucchini
1/3 cup liquid non dairy creamer
3 eggs or egg substitute
1 teaspoon unsalted zero trans fat margarine
1/4 cup brown sugar substitute
1/2 teaspoon all spice
1 1/2 cups plus 2 Tablespoons unbleached flour
2 teaspoons baking powder

Grease pan lightly. Heat to 350 F. Turn when bubbles on pancakes break.

Baked Hash Browns

2 cups frozen hash browns soaked in filtered water for 4 hours and drained
1 1/2 Tablespoons minced onion
1/4 teaspoon minced green pepper
1/8 teaspoon black pepper

Place on sprayed baking dish 9-10 inch. Spray top of potatoes lightly with olive oil spray. Cook on middle rack at 425 for 10-15 minutes. Remove when brown.

Baked Zucchini or Egg Plant

Use either zucchini or egg plant

Wrap vegetables with skin on in aluminum foil. Bake in oven at 300 F until fork tender.
Can be cooled then frozen. Remove foil and reheat in microwave.

Baked Green Beans and Onions

1 lb fresh green beans
1 small onion chopped
1 glove garlic chopped
1 teaspoon bottled lemon juice
1 Tablespoon olive oil

Cook beans in medium saucepan in boiling water with onion & garlic. Add lemon juice and oil. Cook until fork tender.

Honey and Cranberry Acorn Squash

1 small acorn squash peeled and soaked for 4 hours in filtered water and cut in half. Remove seeds with ice cream scoop
!/2 can whole cranberry sauce
3 Tablespoons honey or sugar substitute
2 teaspoons powdered ginger

Cut squash in 1 inch cubes. Steam, or boil in large Dutch oven until just tender. Drain and put in large bowl. In small bowl mix the rest of the ingredients. Pour on squash and toss.

Roasted Italian Spaghetti Squash

1 small spaghetti squash cut in half and seed with ice cream scoop.
2 Tablespoons olive oil
1/2 Tablespoon oregano
1/2 Tablespoon garlic

Place in shallow baking pan and bake for 40 minutes at 375 F shell side down. Scoop out squash with ice cream scoop. Toss with oil and spices.

Pearl Onion Side Dish

1/2 bag pearl onions
1 teaspoon basil
1 1/2 teaspoons olive oil
1/4 cup red pepper

Boil onions for 5 minutes. Slip from skins. Place in bowl and toss with other ingredients.

Roasted Garlic Mashed Potatoes

4 medium potatoes peeled and soaked in filtered water for 4 hours.
1 medium head of roasted garlic
2 Tablespoons zero trans fat margarine
1 teaspoon parsley
2/3 cup liquid non dairy creamer

Place potatoes in saucepan covered with filtered water. Boil gently until fork tender. Drain. Put back in pan. Add rest of ingredients. Mash or use electric mixer.

To roast a garlic head, cut off top and place in oven at 300 until soft, about 30 minutes. Squeeze out garlic cloves. Freeze leftovers.

Roasted Baby Carrots and Rutabaga

1 16 oz package baby carrots cut
1 small rutabaga
1 Tablespoon olive oil
1 small onion sliced
1/2 Tablespoon parsley
1/2 Tablespoon basil
1/8 teaspoon pepper

Peel and dice rutabaga put in medium size bowl; add carrots and the rest of the ingredients. Spray baking sheet with olive oil spray. Arrange vegetables in a single layer. Bake at 450 F until tender, about 25 minutes.

Oriental Peas

1/2 10 oz package frozen peas
1/2 cup reduced sodium chicken broth or stock
8 pearl onions, boiled with skins slipped off
1 teaspoon zero trans fat unsalted margarine
1 Tablespoon corn starch
1/4 cup filtered water
1 small can water chestnuts, drained and rinsed

Combine all ingredients in medium saucepan except corn starch and water. Cook until desired tenderness. Add corn starch to water. Stir until dissolved, add to peas. Heat and stir until thick.

Roasted Three Pepper Zucchini

small zucchinis washed and cut in 1/2 inch thick slices
1/2 cup fresh or frozen onion
3 peppers fresh or frozen
1/4 Tablespoon thyme
1 teaspoon molasses or zero trans fat unsalted margarine
1/4 cup chow mien

Place zucchini and all vegetables in skillet or steamer covered. Cook until desired tenderness. Place in bowl. Add thyme, molasses and noodles.

Toss and serve.

Refrigerator Pickles

4 pickling cucumbers, washed and sliced
1/2 cup white vinegar
1/2 cup filtered water
2 teaspoons pickling spice
2 Tablespoons sugar or sugar substitute
1/2 teaspoon garlic powder

Place all ingredients in small clean glass jar with lid. Shake until sugar dissolves. Refrigerate for 2 days before use.

Apple Stuffed Pork Chops

1 teaspoon lemon juice
4 thick pork chops
1 medium apple washed and diced
1 1/2 cups packaged cole slaw blend
1/2 cup apple cider
4 teaspoons sugar
2 tablespoon sugar free maple syrup
2 teaspoon Dijon mustard

Combine apples, lemon juice, cabbage, cider, syrup and mustard in large sauce pan. Cook 5 minutes covered until cabbage is tender. Cut pocket in pork chops and fill with apple slaw mixture. With olive oil spray coat fry pan. On medium heat cook pork chops 3 minutes each side. Add 1/4 cup apple cider. Reduce to low and cook covered for 8 minutes.

Braised Beef or Pork Short Ribs

1 cups water or 2 cups low sodium beef broth
1 onion sliced
1 clove garlic
4 teaspoons marjoram

Put all ingredients in roasting pan or slow cooker, cover with lid. Bake in oven 35-40 minutes at 350 F or 4 hours in slow cooker. Top with BBQ sauce.

Broiled Steak

Marinate:
4 tablespoons olive oil
1/2 teaspoon cumin
1/2 teaspoon turmeric
1/2 teaspoon garlic powder
1 teaspoon lemon juice
1/2 teaspoon pepper

Tenderize meat with mallet. Place in zip lock bag with all ingredients. Shake until coated. Refrigerate over night; turning bag often. Place foil on broiling pan Broil until meat thermometer reaches 160 F rare, 170 F medium, 180 F well done. Remove let rest 5 minutes, serve with garlic butter.

Chicken and Biscuit Casserole

Biscuit mix:

2 envelopes of rapid rise yeast
1/2 teaspoon sea salt (needed for yeast to rise)
3/4 cup filtered water
1 1/2 cups unbleached flour
3 Tablespoons olive or canola oil

Casserole:

1 cup diced cooked skinless chicken
1/4 cup chopped green peppers
1/4 cup chopped onion
1 teaspoon chives
1/4 cup flour
1/8 teaspoon pepper
2 cups liquid non dairy creamer
1/2 cup canned peas
1/2 cup canned carrots
Olive oil

In a skillet, brown onions, peppers and chives. In a small amount of olive, oil blend in flour and pepper. Gradually add non dairy creamer. Heat in pan until thickened. Add chicken and vegetables. Pour into sprayed casserole dish. Mix biscuit mix and cover casserole with biscuit mix. Bake at 425 F for 12-15 minutes

Chicken Patties

1/4 cup butter
1/4 cup flour
1 cup milk or non dairy creamer
1 teaspoon parsley
1/4 teaspoon turmeric
1/8 teaspoon pepper
1 clove garlic
2 eggs beaten
1 1/2 cups ground chicken or turkey
1 1/2 cups bread crumbs

In skillet melt butter. Blend flour into milk and add seasonings. Add to butter. Heat until thick. Mix in chicken and 1/2 cup bread crumbs. Chill until mixture can be shaped into 8 patties. Dip in egg and shake in plastic bag with bread crumbs. Fry in a little olive oil with a lid.

Easy Parmesan Chicken Breasts

4 large chicken breasts
1 cup mayonnaise
1/2 cup grated parmesan cheese
Bread crumbs
Oregano
Garlic

Mix mayonnaise with cheese. Remove any chicken skin and discard. Spread washed chicken with mayonnaise. Sprinkle with bread crumbs. Sprinkle with garlic powder and oregano to taste. Bake 350 F 45 minutes or until juices run clear.

Florentine Pork Chops

4 boneless pork chops
1/4 cup unbleached flour
1/4 teaspoons thyme
1 teaspoon basil
1/8 teaspoon nutmeg
2 tablespoon oil or canola oil
2 leeks sliced
1/4 cup filtered water
7 oz package of fresh baby spinach (soaked over night)
1/4 cup shredded imitation Swiss cheese
1 teaspoon bottled lemon juice

1/2 bag egg noodles cooked and tossed with 2 tablespoons olive oil
In plastic bag, place flour and spices. Put in chops one at a time and shake. In a large skillet add olive oil over medium heat. Brown chops 2 to 5 minutes. Turn, add leeks. Cook another 3-5 minutes. Add water, lemon juice and spinach. Cover and steam for 15 minutes. Sprinkle with shredded cheese and place on noodles.

Hawaiian Meatballs

1 1/2 lbs lean ground beef
2/3 cup unseasoned bread crumbs
1/2 cup minced onion
1 egg or egg substitute
1/4 teaspoon ginger
1/4 cup liquid non dairy creamer
Sauce:
1 can pineapple tidbits (drain and set juice aside)
1/2 cup brown sugar substitute
1/2 cup vinegar
1 Tablespoon worcestershire sauce
2 Tablespoons cornstarch
1 medium green pepper
1 tablespoon olive oil

Combine meatball ingredients. Bake in oven at 325 F for 15-20 minutes. In large skillet pour pineapple juice, brown sugar, vinegar and corn starch. Add pineapple and peppers. Cook for 1 minute longer.

Honey Mustard Chicken

4 pieces of chicken washed
1/4 cup mayonnaise
1/8 cup honey
1/8 cup brown mustard
parsley

Heat chicken in baking dish at 350 F for 30 minutes. Drain fat. In small bowl mix all ingredients. Put a teaspoon of sauce on each piece. Return to oven for 10-15 minutes. Sprinkle with parsley.

Italian Sausage with Pasta

12 oz noodles cooked as directed
1 lb Italian sausage cut in to 1/2 pieces
1/4 teaspoon cayenne pepper
1 onion diced
1 1/2 teaspoons garlic powder
1 small jar roasted marinated red pepper
6 oz slice water chestnuts
1/2 cup reduced sodium chicken broth or filtered water
3 tablespoons parsley

Cook pasta. In a large skillet place sausage and pepper, cook for 5 minutes. Add onion and garlic and cook for another 5 minutes. Add roasted peppers and 1/4 cup marinade. Cook until sausage is done. Add pasta and 2 tablespoons olive oil and toss.

Oriental Glazed Chicken Bites

2 chicken breasts diced into 2 inch cubes
2/3 cup olive oil
rice
Marinade:
1 egg white
4 teaspoons cornstarch
1 tablespoon honey
Sauce:
1 teaspoon cornstarch
3 tablespoons Mandarin Orange Syrup from can
1 Tablespoon honey
1 Tablespoon brown sugar
1 1/2 teaspoon Chinese five star spice
1 teaspoon rice vinegar
1 can mandarin oranges
1/2 cup fresh or frozen diced red pepper
1/2 teaspoon garlic

In large bowl place chicken in marinade and coat chicken. Cover and refrigerate until ready to cook. Combine in small bowl seasoning sauce. Place olive oil in Wok. Coat chicken with seasoning sauce. Cook 10 minutes. Serve with rice.

Pina Colada Chicken

1 cup pineapple drained
1/4 cup juice from canned mandarin oranges
1/2 cup canned mandarin oranges
1/2 cup pineapple juice
2 1/2 teaspoons corn starch
1/2 teaspoon white vinegar
1 1/2 tablespoon coconut extract
4 chicken pieces
1 cup brown rice

Wash chicken. Place in baking pan or microwave safe dish. If microwaving cook 6-9 minutes on high until chicken is done. In small bowl mix sauce ingredients. Microwave until thick. Pour over chicken and serve with rice.

If using oven place chicken in baking pan. Cook at 350 F until done about 30 minutes. In small pan on stove place all sauce ingredients. Heat until thick. Pour over chicken and serve with rice.

Ravioli in Garlic Butter Sauce

1 package frozen meat ravioli
1 stick zero trans fat margarine
1/2 teaspoon basil
1/2 teaspoon oregano
1 Tablespoon minced bottled garlic
Pepper to taste

Cook ravioli as directed on package in filtered water. In small sauce pan melt butter and add spices. Pour over ravioli and serve hot.

Rice Stuffed Flounder

1/2 cup rice
1 cup broccoli cut in small pieces
1 leek slice
1/4 teaspoon parsley
1 egg yolk
2 Tablespoons brown mustard or honey mustard
3 Tablespoons lite mayonnaise
4 flounder fillets
1/8 teaspoon lemon pepper
1/2 teaspoon bottled lemon juice
1 cup filtered water

In medium saucepan place water, rice, broccoli and leek. Turn on high until boils then turn to low until water is absorbed.
In medium sized bowl place parsley, yolk, mayonnaise, lemon juice, and mustard. Stir well and add rice. With olive oil spray a 13x9 baking pan. Place fish on 1/2 of fish fillet place some of rice mixture. Fold over fillet. Secure with tooth pick. Add 3/4 cup filtered water. Cook at 400 F for 20-25 minutes

Roasted Chicken Sausage

5 red potatoes, peeled, quartered and soaked in filtered water for 4 hours
1 medium onion
2 cloves garlic, peeled
3 medium carrots, cut in slices
1 small green pepper, sliced
1/2 teaspoon olive oil
4 cooked chicken sausages
teaspoon rosemary

In large bowl coat vegetables evenly with rosemary, pepper and oil. Place in large roasting pan and spread out evenly. Bake 15 minutes at 425 F. Add sausages to the vegetables. Cook until potatoes are tender (tossing occasionally), approximately 15 minutes.

Seared Pork Loin and Vegetables

1/4 pork loin
1 teaspoon garlic powder
1 teaspoon coarse black pepper
1/2 teaspoon rosemary
Olive oil
Rub meat with olive oil. In small bowl mix spices. Rub meat with spices. Heat fry pan to high. Put in meat sear on all sides until brown. Turn to low. Add 1 tablespoon olive oil and continue cooking 8 minutes with lid. Turn off.
Vegetables:
1 leek sliced
2 stacks celery sliced
10 sugar snap peas cut up small
1 teaspoon minced dried onion
6 small red potatoes, peeled, cut in quarters and soak in filtered water for 4 hours
3 teaspoons olive oil

Put all vegetables in skillet with marinate oil. Cook 20-25 minutes on medium to medium low. Cover with lid stirring occasionally.

Seared Tuna and Bow Ties

1/2 box of bow tie pasta
1/2 cup broccoli
1/2 cup cauliflower
1/2 cup onions
1 small can tuna in water drained
1 teaspoon olive or canola oil
1/2 teaspoon basil
1/2 teaspoon garlic
1/4 teaspoon pepper
Ranch dressing

Bring large pot of filtered water to boil and add pasta, broccoli, cauliflower and onions. Boil approximately 7-8 minutes or until done. Drain water and add tuna that has been seared in a small fry pan with 1-2 teaspoon olive oil and spices. Serve coated with more olive oil, and add Ranch dressing to coat.

Shrimp with Mixed Peppers

1 large onion diced or 1/2 cup frozen
2 cups peppers
1/2 teaspoon garlic powder
1/2 pound medium shrimp (uncooked and peeled)
1/3 cup bottled lemon juice
1/8 - 1/4 teaspoon cayenne pepper to taste
1 teaspoon cilantro
1 teaspoon basil
1 1/2 tablespoon olive or canola oil

In large fry pan heat oil on medium. Add peppers, onions and spices. Sautee lightly. Add shrimp. Stir often and cover. Cook till shrimp is pink. Serve over rice with a little more oil or pasta noodles of your choice.

Tuna Bake

Biscuit Mix:

1 1/2 cups unbleached flour
2 envelopes rapid rise yeast
1 tablespoon margarine
1/2 teaspoon sea salt
1 cup liquid non dairy creamer
3 tablespoons olive or canola oil
1 egg or egg substitute
1/4 cup sugar or sugar substitute

Mix all together in an olive oil sprayed square baking pan.

In a small bowl place
1 can tuna packed in water drained
2-3 tablespoons low fat mayonnaise
1 teaspoon bottled lemon juice
1/4 cup celery sliced
1 leek sliced
1/8 teaspoon pepper

Place on top of biscuit mix Sprinkle with 1/2 cup unseasoned white bread crumbs mixed with 2 tablespoons melted soft tub or zero trans fat margarine Bake 30 minutes.

Turkey with Fruit Stuffing

2 cups diced and cooked turkey
1/2 to 3/4 bag unseasoned white croutons
1/2 small can lite or no sugar added cherry pie filling
1/2 small can pineapple tidbits (natural juice)
1/4 cup dice onion
1 teaspoon sage
1 teaspoon margarine
1/2 teaspoon pepper
1 1/2 cups filtered water

Spray 8x8 pan with olive oil. Place croutons in pan. Mix with cherry pie filling and pineapple. If not enough use a little more filtered water. Arrange turkey on top. Cover with foil and bake at 350 F for 20-25 minutes.

Swedish Meatballs and Noodles

Meatballs
1 lb lean ground beef
1/2 lb lean ground pork
1/4 cup dry breadcrumbs plain
1/3 cup minced onion
1/8 teaspoon garlic powder
1 teaspoon worchestershire sauce
1/2 teaspoon ground allspice
2 eggs or egg substitute

Place all ingredients in large mixing bowl. Mix well. Shape into balls. Place in a large casserole dish Bake in oven 375 F for 20 minutes drain any grease.

Sauce
2 tablespoons no salt zero trans fat margarine
3 tablespoons unbleached flour
1 cup low sodium beef broth or stock
1 teaspoon dill weed
1 cup liquid non dairy cream

In large fry pan melt margarine and flour. Mix until smooth on low heat. Add rest of ingredients, stir until thick.
1 large bag of egg noodles, boil noodles in filtered water until done
Remove meat balls from casserole dish and set aside. Place noodles drained on bottom of casserole dish. Replace meat balls, cover with sauce.

Apple Butter Cake

2 1/2 cups cake flour
1 1/4 teaspoons baking powder
3/4 cup soft zero trans fat margarine
1 1/2 cups apple butter made with apple juice concentrate
1 cup sugar or equitant equivalent of sugar substitute
2 eggs
1 teaspoon vanilla extract
4 oz cream cheese softened
Powdered sugar

Preheat oven to 350 F. Spray square cake pan with olive oil. In large bowl place margarine and beat until fluffy, approximately 2 minutes. Add sugar and egg, beat 1 minute. Add vanilla extract, cream cheese, flour and baking powder, beat 2 minutes. Pour into pan. Bake 35-40 minutes. Dust with powdered sugar.

Apple Cake

3 cups unbleached flour
1 1/2 teaspoons baking powder
1/2 teaspoon baking soda
1 teaspoon cinnamon
1/2 teaspoon nutmeg
1/2 teaspoon allspice
1/2 cup olive oil
1/2 cup no salt, zero trans fat margarine
1 1/2 cup brown sugar substitute
3 eggs or egg substitute
1 can apple pie filling
Powdered sugar

Place all ingredients in mixing bowl. Beat for 3 1/2 minutes. Pour in Bunt pan. Bake 350 F for about 1 hour. Sprinkle with powdered sugar.

Apple Pie Delight

1 cup margarine
1/2 cup sugar
1 teaspoon cinnamon
1 teaspoon cloves
1 can lite apple pie filling
6 small cooking apples (use 2 different types) peeled and sliced. Coat with 1 tablespoon lemon juice.
1 frozen pie shell
apple cider in season or apple juice

Place margarine, sugar and spices in oven safe skillet. Heat until sugar is dissolved. Add pie filling and apples. Mix until coated. May add 1 teaspoon unbleached flour for thicker juice. Place thawed pie shell on top. Tuck edges with spatula. Brush top with apple cider in season or apple juice. Sprinkle with sugar and cut 4 slits on tip. Bake 30 minutes at 425 F.

Apple Streusel

Cake mix
2 cups cake flour
1 teaspoon baking powder
1 teaspoon baking soda
1/4 cup sugar or sugar substitute
1/4 cup soft tub zero trans fat margarine
1/2 cup liquid non dairy creamer
1 egg
1 teaspoon rum extract

Filling
1 can lite apple pie filling 20 oz
1/3 cup brown sugar or sugar substitute
1/4 cup old fashioned rolled oats
1/4 cup unbleached flour
1/4 cup toasted coconut
3 Tablespoons soft tub zero trans fat margarine melted in microwave
2 teaspoons cinnamon

Mix cake mix. Spray 8x8 baking pan with olive oil spray. Add cake mix.

In medium bowl mix filling, all but apple pie filling. Spread pie filling over cake. With spatula handle poke the apple filling down in cake. Bake 300 F for approximately 30 minutes.

Black Cherry-Vanilla Italian Ice

4 cups filtered water
2 cups sugar or sugar substitute
1 cup frozen black cherries
2 teaspoons vanilla extract

In medium sauce pan place water and sugar, stir until dissolved and boil for 5 minutes. Add vanilla extract. Cut frozen cherries in half and place in 6 single 1/2 cup containers. Pour in sugar mixture and freeze.

Cheese Cake

1 graham cracker crust
1 small box sugar free strawberry gelatin
1/2 cup hot filtered water
8 oz package cream cheese
1 1/2 cup sliced strawberries

In medium size bowl dissolve gelatin in water. With electric mixer add cream cheese. Beat at slow for 2 minutes, then high for 2 minutes. Add strawberries. Mix well. Pour into crust and refrigerate 2 hours.

Cranberry-Maple Pie

3/4 bag fresh cranberries
3/4 cup filtered water
1 pie shell
4 egg or egg substitute
1 cup dark brown sugar or sugar substitute
1/2 cup maple syrup or imitation sugar free
2 Tablespoons rum extract
4 Tablespoons melted zero trans fat margarine in microwave
1 1/2 Tablespoon cornstarch
1 cup mini marshmallows
1 cup white baking chips (optional)
Sugar free whipped cream (optional)

In large sauce pan place cranberries, sugar and 3/4 cup filtered water. Cover and steam on low till cooked halfway. Turn off. Add syrup, extract and cornstarch (mix cornstarch with 1/4 cup cold filtered water). Place marshmallows and chips in pie shell. In cooled cranberry mixture, stir in eggs and pour on top of crust. Bake 350 F for 40 minuets or until center is set. Decorate with whipped cream.

Flan

3/4 cup olive or canola oil
1/2 cup sugar or honey
2 tablespoon vanilla extract
2 eggs
1 1/4 cup unbleached flour
2 Tablespoons baking powder
1/4 cup non dairy creamer
1 package vanilla sugarless pudding
fresh or frozen fruit (optional)

In large mixing bowl add all ingredients. Mix well with electric mixer starting on low. Pour into sprayed flan pan. Bake 20 minutes at 350 F. Cool. Put plate on top of cake then quickly turn over. In the middle place vanilla pudding or top with slices of fresh or frozen fruit.

Frozen Pudding Cream Pie

1 graham cracker pie shell (save plastic for lid)
1 3/4 oz instant sugar free pudding mix
4 package whipped sugar free toping mix
cold non dairy creamer
1 teaspoon vanilla extract
1 teaspoon almond extract
butterscotch or vanilla chips
fresh strawberries

Place pie shell in middle rack of oven. Bake at 300 F for 10 minutes. Cool. In large bowl make pudding according to directions for pie mix using non dairy creamer. Make 2 packages whipped topping with vanilla extract. Fold into pudding, spread into pie shell. Make 2 packages whipped topping with almond extract. Spread on top of pie and freeze. Before serving decorate with butterscotch or vanilla chips. Put fresh strawberries around top, or sprinkles, or additional whipped cream.

Pear Crisp

1/2 teaspoon cinnamon
1 Tablespoon butter or margarine melted
1/4 cup old fashion oatmeal
2 Tablespoons brown sugar
2 cans pears in natural juice drained
1/4 cup brown sugar
1/2 teaspoon clove
3 Tablespoons unbleached flour
non dairy creamer

In small bowl combine oatmeal, brown sugar, cinnamon and butter, mix well. In square baking dish put pears, flour, sugar and cloves, mix until coated well. Sprinkle with oat mixture. Bake 25 minutes at 375 F. Serve with warm non dairy creamer.

Pineapple or Applesauce Spice Cake

Cake
1 1/2 cup unbleached flour
1 cup old fashioned oatmeal
1 1/2 teaspoons baking soda
1 cup packed brown sugar
1 teaspoon cinnamon
1/2 teaspoon ginger
1/2 teaspoon clove
1/2 cup olive or canola oil
1 1/2 cups applesauce or 16 oz canned crushed pineapple drained
2 eggs
3 teaspoons molasses
Toping
1 cup cream cheese
1/4 cup applesauce or 1/4 cup crushed pineapple
1 teaspoon vanilla extract
1/3 cup shredded coconut

Mix all ingredients in medium sized bowl using electric mixer for 3 minutes. Pour into a spray bunt pan. Bake at 350 F for 35-45 minutes or until a knife inserted comes out clean. Mix all toping ingredients well and top cake. Keep refrigerated.

Quick Strawberry Cheese Cake

1 small package sugar free strawberry gelatin
1 graham cracker crust
1 cup boiling filtered water
8 oz softened cream cheese
1/2 teaspoon vanilla extract

Dissolve gelatin in water. Mix in bowl with electric mixer, cheese and vanilla extract. Pour in gelatin and mix well. Pour in crust and chill.

Upside Down Pear Cake

4 tablespoons raspberry preserves
10 oz can natural juice pear halves
1/2 cup unbleached flour
1 envelope sugar free whipped cream
1 teaspoon rum extract
liquid non dairy creamer

Spray lightly 4 microwaveable 12 oz coffee mugs. Place one tablespoon of raspberry preserve on bottom, place pear half on top. Repeat for all cups. Make whipped top as directed with liquid non dairy creamer. Add flour and beat on high for 1 minute. Add rum extract beat 30 seconds. Pour into cups. Place cups two at a time in microwave for 3 minutes on high. Place small plate over cup and turn over. Serve warm.

Manufactured By: RR Donnelley
                   Momence, IL USA
                   December, 2010